Beriah A. Watson

Disease Germs

their origin, nature, and relation to wounds

Beriah A. Watson

Disease Germs
their origin, nature, and relation to wounds

ISBN/EAN: 9783337878047

Printed in Europe, USA, Canada, Australia, Japan

Cover: Foto ©berggeist007 / pixelio.de

More available books at **www.hansebooks.com**

DISEASE GERMS:

THEIR ORIGIN, NATURE, AND RELATION

TO WOUNDS.

BY

B. A. WATSON, M.D.,

JERSEY CITY, N. J.

EXTRACTED FROM THE
TRANSACTIONS OF THE AMERICAN MEDICAL ASSOCIATION.

DISEASE GERMS:

THEIR ORIGIN, NATURE, AND RELATION

TO WOUNDS.

BY

B. A. WATSON, M.D.,

JERSEY CITY, N. J.

EXTRACTED FROM THE
TRANSACTIONS OF THE AMERICAN MEDICAL ASSOCIATION.

PHILADELPHIA:
COLLINS, PRINTER, 705 JAYNE STREET.
1878.

DISEASE GERMS; THEIR ORIGIN, NATURE, AND RELATION TO WOUNDS.

SPONTANEOUS generation is so intimately connected with the germ theory of disease that I shall preface my remarks with a brief *résumé* of the former subject. The ancients supposed that the low forms of life were generated in the matters in which they made their appearance.

Huxley says: "The proposition that life may, and does, proceed from that which has no life, then, was held alike by the philosophers, the poets, and the people, of the most enlightened nations, eighteen hundred years ago; and it remained the accepted doctrine of learned and unlearned Europe, through the Middle Ages, down even to the seventeenth century."[1]

The first spark of light that penetrated this primitive darkness came from an Italian physician, Francisco Redi, in 1668. Redi's experiments proved that maggots are generated not in dead flesh, but from eggs deposited by blow-flies. He was the first to give utterance to this remarkable hypothesis, "no life without antecedent life."

This work, begun by Redi, was continued by Vallisnieri, Swammerdam, and Reaumur, "who succeeded in banishing the notion of spontaneous generation from the scientific minds of their day, and indeed, as regards such complex organisms as those which formed the subject of their researches, the notion was banished forever."[2]

The improved microscope brought to light a world of life in the form of minute organisms; too small even to have been seen by the unaided eye, and their diminutiveness seems to have suggested some mysterious transition of matter into living bodies.

[1] Lay Sermons, Addresses, etc., p. 346.
[2] March 23, 1878, p. 292.

Here arose again the controversy with regard to spontaneous generation.

The affirmative of the question was advocated by Buffon and Needham, the former postulating his "organic molecules," and the latter assuming the existence of a special "vegetative force" which drew the molecules together so as to form living things. The doctrines announced by Needham in 1748 were forcibly opposed by Abbé Spallanzani in 1777, who has shown that if either vegetable or animal infusions are boiled a sufficient length of time, and while boiling hermetically sealed, no form of life will make its appearance until the seal is broken; but if the same infusions are allowed to stand in open vessels they will soon be found to contain numerous infusorial animalcules. He therefore concludes that the germs in the infusions were destroyed by heat.

Schroeder and Dusch have since shown that the germs may be filtered from the air by cotton-wool; it is therefore safe to conclude that the cotton-wool excluded only solid particles, and that these solid particles were the germs from which the infusorial animalcules were generated. More recent investigations by Tyndall, Pasteur, Beale, and others, have shown that the air contains living germs, while an examination of the cotton-wool shows it to have served as a filter. Secondly, Pasteur has proved that these germs were competent to give rise to living forms by simply sowing them in a solution fitted for their development; and thirdly, he showed that the incapacity of air, strained through cotton-wool, to give rise to life, was not due to any occult change effected in the constituents of the air by the wool, by proving that the cotton-wool might be dispensed with altogether and perfectly free access left between the exterior and interior of the experimental flask.

If the neck of the flask is drawn out into a tube and bent downwards; and if after the contained fluid has been carefully boiled, the tube is heated sufficiently to destroy any germs which may be present in the air which enters as the fluid cools, the apparatus may be left to itself any length of time, and no life will appear in the fluid. Although there is free communication between the atmosphere laden with germs and the germless air within the flask, contact between the two takes place only in the tube; and the germs cannot fall upwards, and as there are no currents, they never reach the interior of the flask. It appears that all the early students investigating this subject were of the opinion that boiling a fluid destroyed all living germs; but this opinion is now advocated only by the supporters of heterogenesis.

A writer in the *Medical Record*[1] says: "The verdict in connection with spontaneous generation essentially depends on the answer which can be given to another problem. As the late Prof. Jeffries Wyman said : 'The issue between the advocates and the opponents of the doctrine in question clearly turns on the extent to which it can be proved that living things resist the action of water at a high temperature.'"

It is even now universally admitted that boiling in water, even a few seconds, destroys all living organisms; but we have the highest evidence to show that certain germs resist the action of boiling water several hours, and afterwards germinate readily when placed in suitable soil.

Prof. Tyndall says in regard to the action of heat on seeds : "The botanist knows that different seeds possess different powers of resistance to heat. Some are killed by a momentary exposure to the boiling temperature, whilst others withstand it for several hours. Most of our ordinary seeds are rapidly killed, while Pouchet made known to the Paris Academy of Sciences in 1866, that certain seeds which had been transported in fleeces of wool from Brazil germinated after four hours' boiling. The germs of the air vary as much among themselves as the seeds of the botanist. In some localities the different germs are so tender that boiling for five minutes, or even less, would be sure to destroy them all; in other localities the diffused germs are so obstinate that many hours' boiling would be required to deprive them of their power of germination. . . . The greatest endurance that I have ever observed, and I believe it is the greatest on record, was a case of survival after eight hours' boiling."[2]

Prof. Billroth "discovered the nature and importance of certain glistening spherical bodies frequently found in infusions containing bacteria and called Dauersporen or durable spores by Cohn, although he did not think bacteria were developed from them. Billroth demonstrated that these Dauersporen form micrococci in their interior, which are set free by the bursting of the envelope, and these are capable of multiplication by scission or

---

[1] March 23, 1878, p. 232.
[2] Popular Sci. Monthly, March, 1878, p. 596.

*

of lengthening into bacteria; also that they are endowed with great vitality, and are not destroyed by freezing, boiling, or drying. He had some which germinated after they had been kept dry for eight years; and whenever he wished to make sure of the destruction of the spores contained in his experimental liquids he heated them to 392° F."[1] These recent investigations have thrown much light on a subject where it was greatly needed.

Heretofore experimenters were greatly annoyed by failing to obtain uniform results. It is certainly true that boiling destroys germs to such an extent as to render practicable the preservation of meats, fruits, vegetables, etc., but nevertheless it has often failed in highly putrescible substances. The causes of failure may be readily inferred, and satisfactorily explained by the above-mentioned facts.

Observation and experiment have confirmed the fact that all living bodies take their origin from pre-existing living matter, and further that in every instance where the unaided eye can follow the process of germination the germ is produced from a living parent or parents.

The fish-culture of the present day furnishes a beautiful example of this, and the same may be said of fruit, grain, and grass. In fact, examples of this kind are universal, and no exception to the rule can be cited either in the animal or in the vegetable kingdom.

Is it not therefore logical to suppose that the laws governing the generation of the higher orders of living beings hold also in the creation of the low organisms? Is it rational to imagine that in the former case all forms of life take their origin from living germs, and in the latter to maintain that lifeless matter spontaneously assumes the living state?

Having already considered some of the salient points of the question of spontaneous generation, a knowledge of which enables us to comprehend more fully "The Germ Theory of Disease," I shall now raise the question, What is a germ? The term germ can be properly applied to any particle of living matter possessing the power of germinating when placed under circumstances favorable to this action. These germs vary in size; some being so large as to be readily perceived by the un-

---

[1] Popular Sci. Monthly, February, 1878, p. 401.

aided eye, and others so small as to require the highest powers of the microscope to detect them. Prof. Tyndall has shown by his experiments that even the highest powers of these instruments often fail to indicate the existence of minute particles, germs, etc., although they may be readily seen in the atmosphere by the aid of a beam of sunlight.

Therefore let it be remembered that "the living particle which sprouts from a cell of an adult plant or organism, and is then detached, may be called a germ, as well as the living particle found in the ovum, or the living matter in the ovary from which the new being is evolved. : . . So that *a germ is but a particle of living matter, which has been detached from already existing living matter, and this living matter came from matter of some sort which lived before it.*"[1]

The Germ Theory of Disease presents here for examination the following questions, viz.:—

1st. Do the phenomena of certain diseases depend on the propagation in the system of minute living organisms having no part nor share in its normal economy?

2d. Do these organisms arise from germs?

3d. How do they find their way into the system?

The first question has been carefully studied and thoroughly discussed by many of the most eminent physicians and surgeons of Europe, from both the pathological and biological point of view; and even analogy has been called to its assistance. However, the brief character of this paper does not enable me to present anything beyond a few of the more important facts and discoveries.

Dr. Beale was the first who pointed out the existence in clear and translucent vaccine lymph of minute particles. The existence of these he demonstrated by means of the microscope; and expressed the opinion that the activity of the fluid depended on the presence of these particles.

M. Chauveau experimentally demonstrated that the vaccine virus does consist of minute particles, and his observations have been verified by Dr. Burdon-Sanderson.[2]

Dr. Beale says: "These particles have often been termed *débris,* and have been regarded as quite unimportant elements of the

---

[1] Beale on Disease Germs, etc., p. 10.

[2] Maclagan on Germ Theory of Disease, p. 7.

lymph. To them however the active properties of the lymph are entirely and solely due, and I should be no more inclined, in the absence of the most positive evidence to the contrary, to regard the fluid portion of the vaccine lymph as the active material, than I should be to assume that the fluid in which the spermatozoa were suspended was the fertilizing agent, and that the spermatozoa themselves were merely epithelial *débris* and quite unimportant; or to infer that the fluid in which the yeast fungi or bacteria were growing was the active agent in exciting fermentation, while the actually growing, moving, and multiplying particles were perfectly passive."[1]

Chauveau showed that the active particles subsided after forty-eight hours, and that no effects were produced by inoculating the albuminous supernatant fluid, while the full effects were produced by vaccinating with the deposit.[2]

Do these facts justify the conclusions that the first proposition is proved?

I have satisfied myself by repeating these experiments that the facts are as previously stated.

Dr. L. A. Stimson, writing of these organisms, says: "It is probable that their *rôle*, so far as disease is concerned, is as follows: while they have no power in themselves to excite disease (diphtheria, vaccinia, septicæmia, typhoid fever, etc.), they are able to absorb the poison ('ferment') which is capable of producing it, to 'fix' it, as it is termed, and to give it up to any tissue with which they may come in contact, acting thus as carriers of contagion; then after the abnormal process has been commenced in the body, a change is brought about in the tissues which renders them suitable for the rapid growth and multiplication of the bacteria, which in turn augment the changes in the tissues, and thus there is found a vicious circle, the consequences of which are too often fatal.

"Any agent which destroys the life of the bacteria or prevents their multiplication breaks this circle and renders a cure probable."[3]

Prof. Liebermeister, in discussing the causes of infectious diseases, makes use of the following language: "As an argument in favor of the view that infectious diseases are produced by low

[1] Disease Germs, p. 145. [2] Ibid., p. 146.
[3] Popular Sci. Monthly, No. 34, p. 204.

organisms, it will not be without significance to regard the facts which led in former times to the unexpected acceptance of parasitismus as the cause of disease. I only call attention to the numerous skin diseases produced by fungi, to the trichina disease, to the examples of mycosis intestinalis, which have been observed with increasing frequency in later times, as well as to the development of fungi in numerous other affections.

"Scabies, so long as the itch mite was unknown, was regarded as the prototype of a purely contagious disease, and even after the discovery of the mite there have been endless discussions, until finally this parasite, which is so easily detected, was recognized by all as the sole and satisfactory cause of the affection. The fact that this disease is now stricken out from the list of contagious diseases, and reckoned among the parasitic, shows that we may perhaps expect further changes among infectious diseases.

"In this connection, however, there are facts of considerable importance, which have been furnished by recent investigations into the nature of many contagious diseases in animals and plants. The contagious diseases of the silkworm, which have been a source of so much danger to the silkworm culture, have been proved to be parasitic, and the history of the development of the parasite has been followed pretty thoroughly. In flies and many other insects we have known similar epidemics of a parasitic nature to have taken place.

"The epidemic and contagious diseases of the higher classes of cultivated plants, such as the potato disease, the grape-vine dissease, the ergot of grain, and others, all are derived from fungous growth. The question, too, on which for a long time opinions were divided, as to whether the fungus were the cause, or only the consequence of the disease, has been answered by the botanists with unanimity. Where the development of the fungus had been thoroughly examined, they reported that it was the sole and sufficient cause of the disease. It is clearly evident, too, that the further the progress of investigation advances in human pathology, and the more frequently low organisms are shown in diseases, the more prominently will this question urge an answer."[1]

In several of the contagious diseases new organisms have been observed, and more or less accurately described; although it must be admitted that their exact relation to the disease has not been fully determined in the majority of cases.

It has, however, been satisfactorily shown by Prof. Cohn, of Breslau, and other excellent authorities, that there are several species of these low organisms, and that the bacteria which are formed in decomposing and putrid fluids are different from those observed in disease.

Dr. Sanderson has said : "All microzymes are not contagia, but all contagia may be microzymes."

This brings us to the pertinent question, What is a contagium?

Liebermeister uses the following language : " It is usual *now* to speak of *contagium* as a specific excitant of disease, which originates in the organism suffering from the specific disease; while miasm, on the other hand, is used of a specific excitant of disease, which propagates itself outside of, and disconnected from, a previously diseased organism."[1]

Maclagan says : "A contagium is a morbific agent, which is propagated in, and given off from, the bodies of the sick, and is capable, when received into a susceptible healthy body, of producing in that body a disease similar to the one during whose course it was formed."[2] He further adds : "All we know regarding contagium is that it consists of minute solid particles; that these particles are probably organized; that in chemical composition they so closely resemble the fluids in which they occur, that the chemist fails to detect even their presence, and that they are so very minute that the highest powers of the microscope fail to give us definite information regarding their nature or even their existence."[3]

We will now give our attention to the third question, How do germs find their way into the system? They may gain admission by immediate contact through the agency of carriers, the air we breathe, the drink we take, or the food we eat. These are the recognized methods for the propagation of contagious, miasmatic, and septic diseases; and the Germ Theory seems to possess a special applicability to them; but I shall from this point devote my attention entirely to the relation of this theory to wounds.

---

[1] Ziemssen, Cyclop. Pract. Med., vol. i. p. 25.
[2] Loc. cit., p. 5. [3] Ibid., p. 31.

It will be observed as we progress with this inquiry that the oldest recorded opinion ascribes to the atmosphere a more or less powerful action on wounds. It further appears that this action was supposed to be due to the physical properties of the air—heat, cold, moisture, or dryness.

Hippocrates refers especially to the action of cold in the following: "Cold pinches ulcers, hardens the skin, occasions pain which does not end in suppuration, blackens, produces febrile rigor, convulsions, and tetanus."[1]

The opinion here expressed was firmly and generally maintained by the profession until the early part of the present century. Thus during the first eighteen hundred years of the Christian era we find surgeons almost universally attributing all surgical complications to the action of cold.

The laity, during this period, learned also to regard cold as the greatest enemy of the wounded, and the impression then made on their minds has not yet been fully removed, as is now frequently shown by their anxious question in regard to protecting wounded surfaces. This opinion, although general, was not fully shared by Galen, who, eighteen hundred years in advance of his times, declared that the air often becomes injurious and dangerous in consequence of the heterogeneous substances for which it serves as a vehicle.

Celsus, who had carefully studied wounds, tumors, ulcers, etc., was able only to reproduce the precepts put forth by Hippocrates, and like him recognized the necessity of immediately closing wounds to protect them against the action of cold and heat.

"In the year 160 Galen reproduced the ideas of Hippocrates and Celsus, and, like his predecessors, duly considered the physical qualities of the air. According to him the air might be dangerous by its temperature, its degree of humidity or dryness; he finally added a very important fact, that the air becomes noxious in consequence of the heterogeneous substances it holds in suspension; one should not be surprised by seeing him also attentive to the protection of wounds from contact with the air, and in finding him so enthusiastic for treatment with greasy bodies, or the cerate which bears his name. A distinguished surgeon of the fourteenth century, Guy de Chauliac, busied himself in endeavoring to perfect a system of treatment whereby he might

---

[1] Hippocrates' Works, sec. v. aph. 20.

protect wounds against the physical action of the air. In the sixteenth century we find Ambrose Paré engaged in a similar work, imbued with the same idea, the necessity of protecting the wound against the physical action of the air. Paracelsus, who lived in the middle of the sixteenth century, advised absolute cleanliness of wounds, and the removal of all substances which could possibly irritate them, and is supposed to have admitted the agency of the air as a carrier of foreign substances."

In 1612, Magatus, Professor at the University of Ferare, gave the following rules for the treatment of wounds : "It is necessary," said he, "to avoid with great care: 1st. The contact of the air, because it irritates the wound ; 2d. Movements which might produce derangement of the work of agglutination ; 3d. The removal of the pus, which, far from constituting a bad substance, is the best of topical applications, since nature furnished it."[1]

The views of Magatus were endorsed in England by Wiseman, and in France by Billoste, during the same century.

"In the eighteenth century, Pibrac defended the same principles ; fearing above all the action of cold in wounds, he advises infrequent dressings ; he even proposes, with a view to prevent exposure of the parts to the air, to renew only the external soiled dressings and to leave untouched the charpie which is in immediate contact with the wound."[2]

"J. L. Petit, who lived from 1674 to 1760, and who was one of the great surgeons of his epoch, feared much the action of the air on wounds, and thus he recommended to avoid contact with it when it was possible."[3]

In 1766 McBride published his memoirs on the respective properties of antiseptics, announcing first the precept that the air is the principle which forms the cement or the bond of union of all the elemental parts of bodies ; the preservation of the stability and good state of the body depending on that which prevents the escape of this air. He also professed that the precaution to cover accurately all kinds of solution of continuity has the aim to prevent the escape of air which enters into the constitution of all the parts, and which bears on all the organs, in concert with the other elements of the body. For himself he did not recognize suppuration as due to any other cause than the escape of air, and it is nothing else than incipient putrefaction.

[1] De l'action de l'air sur les Plaies, p. 6.
[2] Ibid., p. 8.
[3] Ibid., p. 9.

Jean Falcon thinks air may prevent the healing of a wound in two ways: sometimes by reason of its quality and sometimes by reason of its substance. He also refers to the drying up of the blood and the gelatinous humors which are the liquids by means of which the union and the agglutination of the lips take place. He further adds that in certain cases the air acts mechanically by preventing a perfect approximation of the lips of the wound.

It will be readily seen from the new theories advanced in regard to the action of the air by McBride and Jean Falcon that they were not satisfied with the correctness of existing ones, but wished for something better.

In 1771 A. Monro, a pupil of Cheselden, became an advocate of the aerophobic ideas of his times. John Hunter declared that the air is not the cause of suppuration in wounds. Some other English authors asserted that wounds suppurate in vacuo, the same as in the open air.

John Bell combated the ideas of Hunter, and regarded the air as possessing irritant properties: he recommended the immediate closure of wounds for the purpose of excluding the air.

During a discussion which took place in the Academy of Surgery in Paris about the year 1825, on the following question: "Apprecier l'influence des choses nommées non-naturelles dans les maladies chirurgicales," the following men attributed to the air an injurious action on wounds: Sancerotte, Dedalot Lafflize, Champeau, Camper, Lombard, and Boyer. Lang was a strong advocate of the aerophobic ideas, but he was opposed by Blondin and Lisfranc.

The illustrious Delpech performed tenotomy subcutaneously for the first time May 19, 1816. The object which he sought to accomplish was the protection of the wound in the tendon from the injurious action arising from contact with the air. He was perfectly successful. His success in this effort aroused in Athens a spirit of emulation, and we find Dupuytren performing subcutaneously myotomy in 1822.

Stromeyer modified and improved Delpech's operation in 1831, and Dieffenbach published in the *Archives générales de Médecine* in 1835 an account of numerous and remarkable successes which he had obtained. Duval, Bouvier, and Jules Guérin contributed to perfect and popularize subcutaneous operations. The latter contended earnestly in favor of the noxious action of the air on wounds, against Malgaigne, Ollivier, and Velpeau, and finally had the satisfaction of partially convincing his opponents of the correctness of his views.

Jules Guérin, in this discussion, brought forward a large number of cases in which he had operated subcutaneously, which had not been followed either by inflammation or suppuration, and in which recovery had been remarkably rapid. He further called attention to the fully admitted fact that in every case of tenotomy which was not performed subcutaneously, the operation was always followed by inflammation and suppuration.

Bouley, Renault, and Bouvier admitted that the influence of the air might be able to retard cicatrization.

" It was reserved for Prof. Bouissan to enlarge on the importance of the action of the air on wounds, and to put forth, in this respect, ideas which contain in brief all the actual doctrine of the nosocomial intoxication."[1] This idea was suggested in the year 1858. In 1861–62 the question of the influence of the air on wounds came up again before the Academy of Medicine on the occasion of a discussion relating to the hygiene of hospitals, during which Piorry, Larry, Duvergie, Gosselin, and Michel Levy expressed their opinions on the action of the nosocomial atmosphere. These honorable academicians no longer considered the nosocomial air as the agent of the inflammation and suppuration of wounds, but they recognized it as capable of determining poisonous accidents either by its absorption directly through the lungs or through the wounds themselves. This atmospheric condition so frequently referred to in these times, and so frequently recognized in overcrowded hospitals, was also known to many of the older physicians. This air of the wards to which Delpech in 1815 had called attention on the occasion of an outbreak of hospital gangrene " has been designated by the name of *nosocomiale malaria*" by Giraldes at this time.

We have attempted to set forth the prevalent opinions, among eminent surgeons from the times of Hippocrates to the present, in regard to the action of air on wounds. We find it now safe to assert that it has been generally admitted, by those most competent to form a correct opinion on this subject, that the air does exert an injurious influence on wounds. In fact, I am not aware that any surgeon of our times is willing to deny it. Certainly

---

[1] De l'action de l'air sur les Plaies, p. 27.

the universal success which has attended the performance of subcutaneous operations, the absence of inflammation and suppuration, the great rapidity with which its wounds heal, and the almost complete absence of all surgical complications, surely speak in language which we are compelled to hear and heed.

Again, the freedom with which we use the invention of Dieulafoy, the impunity with which we pass these needles into visceral organs, cannot fail to teach us an important lesson.

Lastly, let us reflect on the vast difference of success which attends the management of a simple or compound fracture.

The simple or even comminuted fracture is rarely attended with danger to the life of the patient, although in many instances there are extensive subcutaneous lacerations; but so long as the principal bloodvessels are intact, little or no anxiety is felt by the surgeon for the safety of his patient.

How different are the views taken by the surgeon in regard to the safety of his patient in cases of compound fractures, and experience has taught us to be very cautious in regard to our prognosis here.

In fact we have learned that the contact of the air with a wound is always attended with danger.

We recognize it, we admit it, we see it.

How does the air exert its injurious influence on wounds?

We have already the following theories advanced in explanation of its action :—

1st. That it depends on its physical qualities, viz., cold, heat, moisture, and dryness.

2d. That it is due to foreign bodies for which the air serves as a carrier.

3d. That the air acts mechanically by preventing the perfect approximation of the lips in incised wounds.

4th. That the air contains germs which being received into wounds are found to generate living organisms; that these organisms in the wounds give rise to inflammation and suppuration, and even in some cases to the septic diseases which are so frequently found in complications of wounds, especially in overcrowded hospitals and other unfavorable locations.

In regard to the first theory advanced with respect to the action of air on wounds by virtue of its physical properties, it is scarcely necessary even to mention the reasons which probably forced the surrender of the ancient idea, since very little, if any, importance is now attached to it by surgeons.

In fact the practice of surgery of the present day is strongly antagonistic to the idea that cold is greatly to be feared. We now obtain excellent results in the treatment of wounds by the application of cold or ice; the latter should certainly be classed among our best antiseptic agents. The second proposition is undoubtedly correct, although the nature of the foreign bodies was not understood at the time the theory was set forth.

The third theory possesses little practical importance, and it never sufficed to explain the injurious action of the air on wounds in the large majority of cases, or even in any single case satisfactorily when the surgeon had performed his duty well.

The first theory has been proved erroneous; the second is a brief statement of a fact; the third contains very little of importance; and the fourth furnishes the necessary explanation, or, in other words, the fourth theory admits all that is set forth in the second, and at the same time designates in a specific manner the character of the foreign bodies and their *modus operandi*.

Pasteur fully demonstrated that "any liquid, however putrescible in the ordinary condition, may become incorruptible when you have killed all the germs it contains, and when it is protected from contact with those of the air.

"The germs in suspension in the air suffice to provoke the operation of living beings when they are sown in a substance rendered incorruptible. . . . To-day the great majority of learned men have rallied to the doctrine of Pasteur, because the study of germs has not only a theoretical interest, but also considerable practical importance; therefore, thanks to these researches, be they chemical or microscopical, they have enabled us to recognize that among all other detritus there exist, in the atmospheric air, the germs of vegetables and the eggs of animals, infinitely small, all living forms which develop themselves at the expense of their surroundings, and which produce, in nourishing themselves by these elements, the peculiar decompositions which are designated under the name of fermentation and putrefaction. Hygiene and pathology are only benefited by these interesting studies, because a new horizon has been opened before them; it has permitted them to seek the action of these invisible parasites, the cause of the great destructive plagues which have decimated humanity."[1]

1 De l'action de l'air sur les Plaies, p. 82.

It must be generally admitted that the process of putrefaction is accomplished by the growth and multiplication of microscopic vegetable organisms, frequently designated as bacterium termo, although the multiplicity of names applied to these and similar organisms has certainly led to much confusion.

While engaged in the consideration of these organisms in connection with wounds and the septic conditions which are found so frequently complicating them, I desire to call your attention to the following by F. Steudener.

"Pyæmia and septicæmia take the first place among the few infectious diseases in which a vegetable organism, as a living virus, has been examined and demonstrated in a scientific manner.

"To Klebs belongs the honor of having discovered this organism and of having followed out with the greatest accuracy the manner of its propagation and influence, and also of having proven by his experiments the correctness of this view. . . Klebs found, by the examination of the secretions of wounds, vegetable organisms in varying quantities in the thick creamy pus as well as in the ichorous; being however extremely numerous in the latter, and never entirely absent in the former. Klebs found, by further examination, these organisms in the form of zooglen settled on the granulation tissues and ulcerating cartilages. He followed their entrance into the intra-cellular space of the connective tissue, where they excite inflammation and suppuration; the same results are also produced by their entrance into the medullary substance in traumatic osteomyelitis. Here he observed that destructive influence on the vessels from which, in consequence of the penetration of the walls, there is formed an adherent or obstructing thrombus."[1]

Again, in regard to these organisms, it is known that filtration of putrid fluid through porous porcelain under pressure deprives it simultaneously of its offensive smell, poisonous action, and power of generating bacteria; thus may the most virulent septic liquid be deprived of its poisonous properties.

I also fully agree with Prof. William Roberts, who says: "We know further from the evidence I have laid before you, that decomposition cannot take place without bacteria, and that bacteria are never produced spontaneously, but originate invariably from germs derived from the surrounding media. . . . We should

---

[1] Sammlung klinisch. Vorträge, No. 38, S. 300.

probably differ less about antiseptic treatment if we took a
broader view of its principle. We are apt to confound the principle of the treatment with Lister's method of carrying it out.
The essence of the principle, it appears to me, is not exactly
to protect the wound from the septic organism, but to *defend the
patient against the septic poison*."[1]

The diseases arising from septic poison may be truly said to
be the opprobrium of surgery, and further that the chief success
in the management of wounds has always depended, and must
always continue to depend, principally on the ability of the surgeon to protect his patient against this poison. Let us now examine the *successful treatment of thirty-five consecutive amputations
one hundred years ago* by Alanson.

The dressings were changed no oftener than was necessary to
secure perfect cleanliness. In a word, the surgeon with great care
and nicety aimed at primary union; and his success was all that
could be desired. It is also said, " Alanson was fully alive to
the dangers of foul air of hospitals, as shown by the fact that one
patient was removed from the infirmary during a severe attack
of erysipelas, and, after being bathed and supplied with new
clothing, was placed in a building near at hand."[2]

Here we find, more than one hundred years ago, a surgeon
applying the antiseptic treatment, and fully aware of the danger
of exposing a wounded surface to the vitiated air of a hospital.

The only advantages to be gained by what is called The Open
Method of Treatment of Amputations is wholly due to a diminished liability on the part of the patient to be contaminated with
the septic poison, and the phrase "diminished liability" must be
here received in a limited sense. While it may be an improve-
ment over the ordinary closed method, it is certainly far from
being perfect. The open method may succeed, will succeed in a
comparatively pure atmosphere, but in an overcrowded hospital,
where the air is badly vitiated, it should certainly give way to
some better system. Let us glance at the advantages of this
open method.

1st. In this method the probability of *mechanically sealing up
putrescible fluids within the freshly cut surfaces of the flaps, as is
often done by the closed method, is avoided.*

2d. Since nearly all secondary wound complications depend

[1] Med. Times and Gazette, Aug. 11, 1877, p. 138.
[2] Med. Record, vol. ii. p. 756.

on the absorption of putrefactive substances, or their agents, it follows that in the same ratio you prevent this absorption you avoid the appearance of these diseases.

I will also refer to a favorite dressing in France, "*The Cotton Wadding*," which has received the endorsement of the Paris Academy of Surgery. This method of treatment is based on the idea that the air contains germs which may be kept from coming in contact with the wound, and by these means the patient may receive the required protection.

The Application of Ice in cases of severe contuso-lacerated wounds is a favorite remedial agent with many surgeons. Billroth speaks very highly of it. He applies it in cases of injuries of the extremities, not only to the wound itself, but along the limb above and below it. He allows the ice to remain continuously applied until the laceration has healed. He informs us that the wounds treated in this manner *do not suppurate*, that they are very rarely attended with cellulitis, or any other indication of septic infection. The explanation of these facts is comparatively easy by the aid of the Germ Theory.

We know that all living organisms are killed or rendered torpid by exposure to cold, and that germs do not germinate in ice. For these reasons we may expect a certain degree of benefit to be obtained by the industrious use of cold water, but the direct and continuous application of ice or ice-bags will be found much more advantageous in a majority of cases. There are some cases of amputations and other wounds which may be successfully treated by irrigation, providing proper caution is had to make it thorough and absolutely continuous. I would suggest, however, that the water used should contain antiseptics, although the force with which the fluid is thrown on the wound will tend very strongly to wash away organisms and germs.

The Closed Treatment of Amputations.—By this method the flaps are closed after the performance of the operation, or, more correctly speaking, immediately after the cessation of oozing, although some surgeons favor leaving it open a little longer to allow glazing to take place, which is a condition depending on the drying of an exudation on the surface of the wound. It will be observed that Alanson, *even by this method*, obtained excellent results, but it should be remembered at the same time that he made the treatment, by extraordinary care, really antiseptic, at least as nearly so as human skill and attention could under the circumstances. I am, however, here inclined to think that even he had fewer obstacles to overcome in his time than the surgeons of the present day. His operations were performed prior to the use of anæsthetics, and consequently he *probably found less trouble* in completely arresting hemorrhage than we now contend against. The violent struggling of the patient during the operation would cause all vessels to bleed instantly that would bleed at all, and consequently the surgeon would arrest the hemorrhage before leaving his patient.

In this method of treatment the danger to which I desire to especially call attention is the oozing which occurs after the closure of the flaps. We now operate always under the influence of anæsthetics; the principal arteries are promptly ligated; the tourniquet is then removed; if any arterial branch should be found to bleed it will receive promptly the surgeon's attention, but the anæsthetized condition of the patient does not favor hemorrhage. The flaps are closed; reaction commences after the lapse of a longer or shorter period; there is oozing within the closed flaps; the circumstances are in the highest degree favorable to septic poisoning; there is a putrescible fluid, atmospheric air loaded with germs, a proper degree of heat, and, to aid these conditions still further, pressure from without. I question whether human skill can devise more favorable conditions for the speedy germination and certain introduction of this septic poison. It might very properly be mentioned here that there are other important objections to and defects in this method of treatment; but I will merely add that it is certainly not the best method of treatment which can be adopted in the vitiated air of a hospital.

Antiseptic Treatment of Wounds.—The first publication in regard to this treatment was made by Prof. Joseph Lister, in the *Lancet*, March 16th and 23d, 1867. It is not necessary for my purpose that I should here enter into a detailed report on the materials used, and their mode of application, but as Assistant Surgeon A. C. Girard, United States Army, has said, "The only thing which concerns us here is the indisputable fact that there are germs or ferments in the atmosphere which will produce putrefaction in wounds, and that by preventing their ingress we can in most cases avert the complications which cause the greatest fatality in surgery. This is the key to Lister's system."[1]

[1] Surg. Gen. Office Circular Order, No. 3, Aug. 20, 1877.

I feel that it is impossible for me to bring the merits of this treatment before you in a more forcible manner than has been done by Dr. Girard in his excellent report to which I have already alluded, and I only regret that it has not been in the hands of every surgeon in this country. I am thoroughly convinced that he who reads the arguments in favor of this system as they are detailed to-day in the German and French literature cannot fail to be convinced of its value, unless he refuses to be convinced by any argument which he *finds* himself unable to refute.

Let us now further examine Dr. Girard's report. He says: "During a sojourn abroad last winter my attention was particularly drawn to this innovation in surgery, as it has been introduced on the European continent but two years, and was the almost exclusive topic of conversation of the surgical profession there. It happened that my first intercourse was with some of the most decided and renowned opponents of the system, and I became acquainted with all the objections to it before I had witnessed its advantages and benefits.

"I received therefore the glowing accounts of Lister's disciples with an incredulous ear, and it was only by travelling from one 'Lister hospital' to another that belief in its superiority forced itself upon me. I became convinced that if it is not the only proper wound-treatment, it is the safest one, and renders conservative surgery possible beyond what had ever been believed. It would take volumes to describe all that I witnessed, and I cite but a few examples. Who, before this, would have fearlessly opened the knee-joint for suppurative arthritis, as I saw done under the 'spray,' the patient recovering in a few days with a sound joint? Who would have expected an ovariotomy with general adhesions in a woman of seventy-five to heal in eight days without a symptom of reaction, or a laparotomy for the liberation of incarcerated peritoneal hernia in a moribund patient, healing in six days, or a resection of the ulna in nine days? . . . Hospitals which had been in use for centuries, and had become hot-beds of infection, where the majority of operations formerly were followed by pyæmia, gangrene, and erysipelas, where everything had been tried to combat these evils, where treatment 'open,' 'occlusion,' by 'immersion,' compresses of chlorine water, carbolized water, even Lister's 'gauze,' and 'paste' had failed, became entirely free from these complications as soon as Lister's system with *all* its precautions had been introduced. Prof. V. Nussbaum, Surgeon

General in the Bavarian army, told me that formerly he operated in his hospital with the greatest reluctance, as nearly every case was sure to be followed by grave accidents, even the opening of a panaritium or the amputation of a finger would cause pyæmia and death; wounds, granulating in the most healthy manner, as soon as brought into his hospital would become gangrenous, and the patient would die, when a few days before he appeared to be on the eve of entire recovery. Now everything is changed. While during sixteen years in which he had charge of the Munich General Hospital, pyæmia never failed a single month to make its appearance, until at last it seized 80 per cent. of the patients, since the introduction of Lister's system it has absolutely disappeared. The same is the experience of Prof. Volkmann in Halle."[1]

The limited space of this paper does not allow me to enter more fully on the subject of Lister's treatment of wounds, neither do the objects which I have here sought to accomplish require it. My principal aim has been to call the attention of the surgical profession especially to these facts.

1st. That there are certain germs in the air, more particularly in the atmosphere of overcrowded hospitals, which if permitted to enter wounds give rise directly to living organisms, inflammation and suppuration; and indirectly to all septic conditions which are found as wound-complications.

2d. That the successful management of wounds depends principally on the ability of the surgeon to keep the wounds, under all circumstances and at all times, free from germs and living organisms; and therefore the value of any method of wound treatment depends primarily on the degree of antisepsis which can be obtained by it.

3d. That the occasional discovery of a few bacteria in a wound which has been treated antiseptically does not disprove the fact that these bacteria arise from germs; but may be satisfactorily explained in a variety of ways, especially by the existence of germs which have not been destroyed by the means employed.

[1] Surg. Gen. Office Circular Order, No. 3, Aug. 20, 1877, p. 2.

www.ingramcontent.com/pod-product-compliance
Lightning Source LLC
Chambersburg PA
CBHW022034190326
41519CB00010B/1708